Table of Contents

Before the Constitution

Most colonists who settled in America during its first 150 years were British. They worked hard and suffered many hardships, but they persevered as they built settlements along the country's east coast from what is now Maine to Georgia. They lived under the rule of the British.

The colonists asked for a larger role in making decisions affecting them, but the British only tightened their control. By the mid 1770's, the British government had imposed heavy taxes and restricted the colonists' freedom.

On June 7, 1776, Richard Henry Lee, a delegate to the Second Continental Congress, presented the idea that the colonies should be free and independent states. As a result of his presentation, a committee was appointed to write the Declaration of Independence.

Circle the names of the thirteen colonies in the puzzle below. They may be forward, backward, up, down or diagonal.

S	I	O	N	I	L	L	I	N	O	T	G	N	I	H	S	A	W
I	D	A	H	O	R	E	G	O	N	I	N	D	I	A	N	A	E
F	L	O	R	I	D	A	D	N	A	L	S	I	E	D	O	H	R
P	A	N	I	L	O	R	A	C	H	T	U	O	S	M	R	Y	I
A	I	C	C	O	N	N	E	C	T	I	C	U	T	A	T	O	H
I	N	A	N	G	R	I	V	E	R	M	O	N	T	I	H	K	S
P	A	L	E	R	I	H	S	P	M	A	H	W	E	N	C	E	A
P	V	I	W	M	W	E	N	E	V	A	D	A	S	E	A	N	I
I	L	F	J	O	A	D	E	L	A	E	N	T	U	E	R	T	G
S	Y	O	E	N	I	S	N	S	L	K	E	E	H	S	O	U	R
I	S	N	R	A	I	H	W	A	L	O	W	A	C	E	L	K	O
S	N	I	S	N	G	A	W	E	L	Y	Y	S	A	N	I	Y	E
S	N	R	E	T	N	A	A	A	H	Y	O	X	S	S	N	C	G
S	E	A	Y	A	R	Y	W	Y	R	A	R	H	S	N	A	E	N
I	P	V	R	E	I	G	E	O	M	G	K	A	A	E	O	W	J
M	W	W	S	T	V	I	R	G	I	N	I	A	M	T	R	S	Y

Answer the questions below.

1. Why do you think the 13 colonies wanted to be free and independent?

____

____

____

2. Why do you think Great Britain wanted to control the colonies?

____

____

____

FOR EXTRA FUN:

Write ten clues about one of the colonies. Read them to a friend and see if he/she can guess the colony being described.

Leading to the Declaration of Independence

A series of events led to the writing of the Declaration of Independence. Ten years before it was written, the people of America asked the King of England to allow them to have a more active part in making the laws that governed the nation. Their request was denied. England not only continued to make the laws, but they levied more and higher taxes. One tax in particular greatly angered the colonists. This was the Stamp Act. The Stamp Act required the colonists to pay taxes on legal and business papers, newspapers and various other items.

Another tax which made American colonists mad was the tax on many imported goods including tea. It was from the tea tax that the Boston Tea Party was born.

To try to fight against the taxes, delegates from all the colonies except one met in Philadelphia at the First Continental Congress. The delegates made an agreement not to trade with England or to use British goods. But England did not change its ways. A final request asking England to change was sent to King George III just before the Declaration of Independence was written.

Answer the questions and fill in the blanks on page 5 to learn which colony did not agree to stop trading with England. Its name will appear down in the boxes.

1. Write the name of the king who levied high taxes.
2. How many years before the Declaration was written did America ask to be part of the lawmaking process?
3. Where was a "party" held?
4. Goods sent in from another country are called ________ .
5. With what country did America want to share the lawmaking process?
6. Which Continental Congress voted not to trade with England?
7. What act taxed many legal and business papers?

1. [ ] ___ ___ ___ ___ ___

2. ___ [ ] ___

3. ___ [ ] ___ ___ ___ ___

4. ___ ___ ___ ___ [ ] ___ ___ ___

5. ___ ___ [ ] ___ ___ ___ ___

6. ___ [ ] ___ ___ ___

7. ___ ___ [ ] ___ ___   ___ ___ ___

Write a poem about the Boston Tea Party below.

____

____

____

____

____

____

____

Declaration of Independence

The Declaration of Independence was written after the start of the Revolutionary War. This war lasted from 1775-1783 and was fought between the colonies and Great Britain for the colonies' independence. The Declaration of Independence was written in 1776, and it declared America's freedom from British rule. The Second Continental Congress appointed five men, Thomas Jefferson, John Adams, Benjamin Franklin, Robert Livingston and Roger Sherman, to write it. Jefferson wrote the first draft which was almost perfect. The committee made few corrections before it was presented to the Continental Congress. A few more corrections were made, but the document is basically the work of Thomas Jefferson. Church bells rang to signal the adoption of the Declaration of Independence on July 4, 1776—the birth of our nation.

There are five parts in the Declaration: the Preamble, Statement of Human Rights, Charges Against the King and Parliament, Statement of Separation and the Signatures. Though their main purpose was to state the reasons for separation from England, they also expressed some ideals such as all men are created equal and entitled to the unalienable rights of life, liberty and the pursuit of happiness. These ideals are also expressed in the Constitution and are the base today of America's beliefs.

Fill in the puzzle below. The circled letters, when unscrambled, will tell you how many people signed the Declaration of Independence.

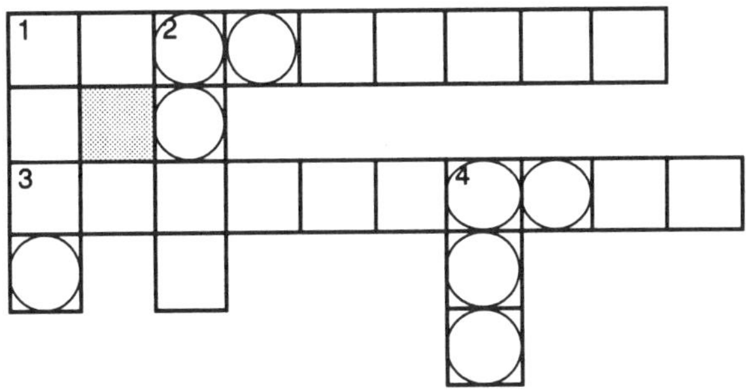

Across

1. The main author
3. A member of the committee who wrote the Declaration

Down

1. The month in which it was completed
2. The parts in the Declaration and the number of people on the committee who wrote it
4. 177__ __ __ , the year it was completed

List ten people who signed the Declaration of Independence.

1. ____________________ 6. ____________________
2. ____________________ 7. ____________________
3. ____________________ 8. ____________________
4. ____________________ 9. ____________________
5. ____________________ 10. ____________________

In three sentences, write why you think no women signed the Declaration of Independence.

____

____

____

____

____

____

Articles of Confederation

During the Revolutionary War, the Continental Congress wrote the Articles of Confederation that were meant to form a national government. The Articles are considered the nation's first constitution. They reflected the distrust that many of the colonists had for a powerful government. Under the Articles, the thirteen colonies became thirteen states with each state maintaining its independence. The states acted like individual countries. They did not work together. They did not want to give up their individual rights as they had under British rule. Each state wanted its own power. The people did not want a strong central government.

The Articles of Confederation set up a Congress of the Confederation to run the government. Some states had more members than other states, but each state was allowed only one vote. There was no President, and there were no courts. Congress was limited to what it could do, and the states did not have to obey the laws it did pass. In 1787, the Constitutional Convention met in Philadelphia and wrote the United States Constitution. It was at this time that the Articles of Confederation were abandoned.

Write **True** or **False** in front of the following statements.

______ The Articles of Confederation formed a strong central government.

______ Every state had one vote in Congress no matter what number of delegates represented it.

______ When Congress asked the states for money to pay national debts, the states did what Congress asked.

______ The President ran the Confederation.

______ Every state did what it wanted.

______ Every state had the same number of delegates in Congress.

______ The courts helped make the laws.

______ The Articles of Confederation was America's first constitution.

______ Every state had to send men to serve in the army when laws were passed establishing an army.

______ The President was elected by the Congress.

______ The Continental Congress wrote the Articles during the Revolutionary War.

______ The Articles of Confederation worked well.

Write three ways you think the Articles of Confederation could have worked better. Be specific.

____

____

____

____

____

____

____

____

Why the Constitution Was Written

Because the thirteen states acted as thirteen separate nations under the Articles of Confederation, the United States was not a unified country, and its government was considered weak. Many leaders of the time recognized the need for a stronger government. In order to establish this, delegates from five states—New York, New Jersey, Virginia, Pennsylvania and Delaware—met in Annapolis in September of 1786 at what is known as the Annapolis Convention.

At the Annapolis Convention, the delegates discussed changes that could be made in the Articles of Confederation. They discussed the trade and boundary problems between states and the states' rights to do as they pleased under the laws of the Articles. From that meeting, a recommendation was made that a special convention be held in Philadelphia in 1787 to revise the Articles of Confederation. Delegates from twelve states met in May of 1787. One state refused to send delegates because it was afraid it would lose its individual rights as a state. When the delegates began to discuss ways of correcting the Articles, they realized they would have to write a new document to overcome its faults. And so they worked for four months behind closed doors writing the Constitution.

Answer the questions. Then, unscramble the circled letters to learn which state did not come to the Constitutional Convention.

What was the name of the convention held in September, 1786?

_ _ ○ _ _ ○ _ ○ _

In what city was the Constitution written?

_ ○ _ _ _ ○ _ ○ _ _ _ _

How long did it take to write the Constitution?

_ _ _ ○   _ _ _ _ _ _ ○

What kind of problems between states did the Annapolis Convention discuss?

_ _ _ _ ○   ○ _ _   _ _ _ _ _ ○ _ _ _

What was the name of the state that did not attend the Constitutional Convention?

_ _ _ _ _   _ _ _ _ _ _

On the map below, label the states which sent delegates to the Constitutional Convention. Put an X on the state which did not send delegates.

Getting Started

George Washington had won the respect of the delegates as commander of the Revolutionary War Army and was elected president of the Constitutional Convention at its opening on May 25th, 1787 in Pennsylvania's State House. (The convention was supposed to have opened on May 14, but since few of the delegates had arrived by that time, it was pushed back to the 25th.) As president, Washington did not make many contributions at the convention. However, he did keep the meetings orderly and running as smoothly as possible considering the many different points of view that were represented by the 55 delegates that attended the convention. Any remarks Washington did make were listened to attentively by the delegates.

The delegates who attended the convention came from twelve of the thirteen states. Rhode Island refused to send delegates as it did not want national government interference in its affairs.

Before the convention got underway, a rules committee organized the procedures the convention would follow. Each state was given one vote. If there was more than one representative from a state, the delegates would have to decide how to cast their one vote. What happened at the meetings was to be kept secret until the entire Constitution could be presented to the public. Any delegate could voice an opinion. A vote cast one way could be changed if the delegates saw it was necessary as they proceeded.

Complete the crossword puzzle below.

Across

1. War Washington commanded
5. Elected Washington president of convention
7. Where convention took place

Down

2. Means by which different laws were settled
3. Month in which the convention began
4. Way in which meetings were held
6. Number of votes allowed per state

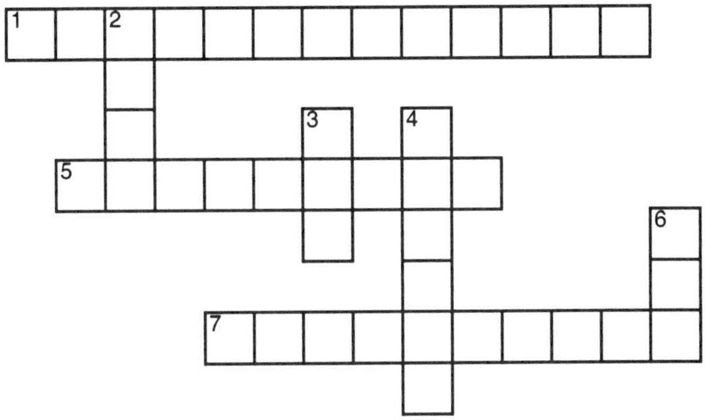

Choose a historic person who lived at the time the Constitution was being written to write five clues about. Read the clues to the class to see if they can guess who they describe.

The Great Compromise

As the Constitution was being written, several areas of disagreement became apparent. The area causing the greatest amount of concern was the number of representatives each state should be allowed to have in the legislature. The larger states thought representation should be determined by a state's population. This was called the Virginia Plan. The states with smaller populations were not in favor of such a plan. They wanted all states to have an equal number of representatives. Their plan was called the New Jersey Plan. Roger Sherman of Connecticut proposed a two-house legislature. In one house, the Senate, there would be an equal number of representatives. The other, the House of Representatives, would have a representative for every 30,000 residents. This plan satisfied the large and small states and became known as the Great Compromise.

The number of representatives each state should have was not the only area in which delegates disagreed. Compromises also had to be made concerning the issue of slavery and concerning how to count slaves to determine the number of congressmen a state could have.

Using words found in the information above, write another word for each of the following:

argument ________________________________

identical ________________________________

give-and-take ________________________________

sundry ________________________________

affair ________________________________

Circle the answers in the puzzle and write them after each question. The letters that remain uncircled in the puzzle will spell first, the name of Virginia's governor and the man who presented the Virginia Plan, and second, the man who presented the New Jersey Plan.

A	I	N	I	G	R	I	V
E	S	E	D	M	U	R	L
N	H	W	D	R	E	L	A
E	E	J	N	G	A	D	O
T	R	E	O	M	L	P	L
A	M	R	S	H	W	A	I
N	A	S	L	L	R	I	A
E	N	E	M	G	P	A	T
S	E	Y	E	R	S	O	N

1. Which states favored the New Jersey Plan? ________________
2. Representation based on population was which plan?

   ____
3. What the first name of the man proposing the Great Compromise? ________________
4. What was his last name?

   ____
5. Which plan suggested equal representation? ________________
6. Which legislative body has an equal number of representatives? ________________
7. Which states favored the Virginia Plan? ________________

   Which man proposed the Virginia Plan? ________________

   Which man proposed the New Jersey Plan? ________________

Write about a current situation in the news in which a compromise is needed. Write what you think would be a viable compromise and tell why.

____

____

____

____

____

____

____

The Three-Fifths Compromise

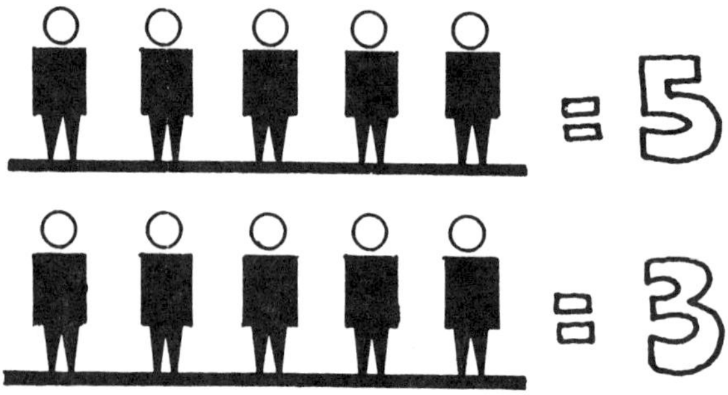

The Northern states were opposed to slavery, and the Southern states were not. The Southern states wanted their slaves counted in the population when it came to the number of congressmen that were allowed in the House of Representatives. They did not want them counted when figuring a state's taxes based on the number of its residents. The Northern states disagreed. The framers of the Constitution never used the words slave or slavery. They suggested that all free persons and three out of five of "all other persons" be counted in the population for the purposes of taxation and representation in Congress. This became known as the Three-fifths Compromise.

Give some thought to the questions below and on page 17 and then write your answers.

What does it mean to compromise? ____________________

____

Why were the Southern states willing to count every slave? _____

____

Why did the North want every slave to be counted? __________

____

Where was this compromise originally stated in the Constitution?

____

Explain how this compromise worked. ________________

____

____

____

Do you think it was a fair compromise? _______ Why or why not?

Find and write about another famous compromise involving the U.S. ___

Write about a compromise in which you have been involved. Do you think it was fair? _________ Why or why not? ___

Do you think compromise is a good way to solve disagreements? _________ Why or why not? ___

Signing the Constitution

Only about thirty-five of the delegates were present for most of the entire proceedings at the Constitutional Convention. Seldom were all fifty-five there together. The New Hampshire delegates arrived nine weeks late. Other delegates had to leave to take care of business at home. Four delegates, Luther Martin, Robert Yates, John Lansing, Jr. and John Francis Mercer, disapproved of the Constitution and were absent for its signing. Nine delegates, Oliver Ellsworth, Caleb Strong, William Pierce, William Houstoun, Alexander Martin, William Richardson Davie, William Churchill Houston, George Wythe and James McClurg, approved of the Constitution but were absent at the time of its signing. John Dickinson could not be present for the signing, but he arranged for a fellow delegate to sign his name. Edmund Randolph, George Mason and Elbridge Gerry were not absent, but they did not sign the Constitution. Six delegates, Nicholas Gilman, Jared Ingersoll, Richard Bassett, John Blair, William Few and William Blount, always voted for issues when they were there, but never said a word during the proceedings.

Listed below and on page 19 are the names of the fifty-five delegates to the Constitutional Convention. Circle the names of those that signed the Constitution.

New Hampshire
John Langdon
Nicholas Gilman

New York
Robert Yates
Alexander Hamilton
John Lansing, Jr.

Massachusetts
Elbridge Gerry
Nathaniel Gorham
Rufus King
Caleb Strong

New Jersey
David Brearley
William Churchill Houston
William Paterson
William Livingston
Jonathan Dayton

Connecticut
William Samuel Johnson
Roger Sherman
Oliver Ellsworth

Georgia
William Few
Abraham Baldwin
William Pierce
William Houstoun

Virginia
George Washington
Edmund Randolph
John Blair
James Madison, Jr.
George Mason
George Wythe
James McClurg

North Carolina
Alexander Martin
William Richardson Davie
Richard Dobbs Spaight
William Blount
Hugh Williamson

Delaware
George Read
Gunning Bedford, Jr.
John Dickinson
Richard Bassett
Jacob Broom

South Carolina
John Rutledge
Charles Pinckney
Charles Cotesworth Pinckney
Pierce Butler

Maryland
James McHenry
Daniel of St. Thomas Jenifer
Daniel Carroll
John Francis Mercer
Luther Martin

Pennsylvania
Thomas Mifflin
Robert Morris
George Clymer
Jared Ingersoll
Thomas FitzSimons
James Wilson
Gouverneur Morris
Benjamin Franklin

How many delegates signed the Constitution? __________

There was actually one more person who signed it. His name was William Jackson. He was secretary of the Convention. How many actual signers were there? __________

In which states did all the delegates sign the Constitution? _____

____

In which state did only one delegate sign? ____________________

How many delegates actually did not approve the Constitution?

____

Which states had the most delegates that did not approve it?

____

Which state had the largest delegation? ____________________

Which had the smallest? ____________________ What did they have in common?____

Fathers of Our Country

The fifty-five delegates who met in the summer of 1787 in Philadelphia had been selected by their states. They were educated, patriotic and experienced men. Over half of them were lawyers and judges, a fourth were landowners, all of them had held at least one public office, and all of them had backgrounds of some financial success. Many of the delegates were under fifty years of age. Benjamin Franklin was the oldest at 81. Though there were many differences between the delegates' ideas, they were a daring, courageous and creative group of men who were willing to take steps to establish a stronger national government. They debated, wrote and rewrote the Constitution during the summer of 1787. Although all the delegates were not always present, they had the opportunity for input into the document that is now the law of our land. Because the Constitution is the basis for our country and the product of their hard work, the men who wrote it have been named the Fathers of Our Country.

Write **True** or **False** in front of the statements below and on page 21.

_______ Many of the delegates' backgrounds were similar.

_______ Only the men who were present all the time are the Fathers of Our Country.

_______ Many of the delegates were young.

_______ The delegates were young and inexperienced.

_______ The delegates disagreed on many of the ideas in the Constitution.

_______ No candidate was older than 81.

_______ The purpose of writing the Constitution was to make laws.

_______ Only special candidates were allowed to present ideas.

_______ Only one draft of the Constitution was made.

_______ Anyone who wanted to be a delegate could come to the Constitutional Convention.

_______ The Constitution presents laws for running our country.

_______ Benjamin Franklin was too old to be a delegate.

_______ Less than fifty-five men wrote the Constitution.

_______ Writing the Constitution was not easy.

Rewrite the false statements to make them true.

Reading the Constitution

Look at a copy of the entire Constitution. It is only 4,300 words long. After you have looked at it, answer the questions or fill in the blanks below and on page 23.

The Constitution is divided into _____ main parts.

They are the ____________, ____________ and ______________.

What is the purpose of the Preamble? ______________________

____

____

There are _____ articles in the Constitution. Many of them have several sections. Tell which article provides for the following services or laws.

- United States court system ______________
- The nation's debts and upholding the Constitution __________
- The lawmaking body or the government __________
- What must be done for the Constitution to be law __________ (This process was called __________________ .)
- Explains the duties of the President ______________
- Allows for changes to be made in the Constitution __________ (That process is called __________________ .)
- Tells what states and federal government can do __________

There are _____ amendments in the Constitution.

What makes the Constitution a usable document today? ______

____

____

____

____

Which article has allowed it to be an up-to-date document?

____

Define the following words as they relate to the Constitution.

ARTICLE ____

AMENDMENT ____

RATIFY ____

Choose the amendment you think is most important. Write it below. ____

Why do you feel this is the most important amendment? ______

____

____

____

____

Write about a situation, real or pretend, involving the amendment. ____

____

____

____

____

____

____

____

____

____

____

____

____

____

____

Three Branches of Government

The Founding Fathers did not want all the powers given to the government to be controlled by one man, or even just a few men. They feared that if a small group was given too much power, the United States would once again be ruled by a tyrannical government like it had been under England. To avoid such a situation, they divided the government into three branches: the executive, the legislative and the judicial. (These three branches are explained in the first three of the Constitution's seven articles.)

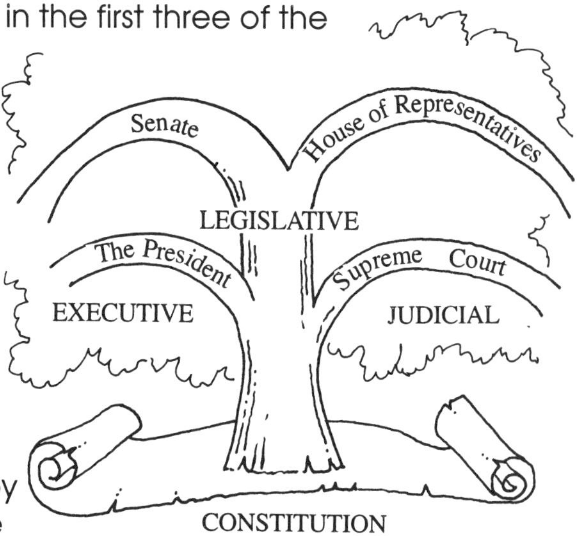

Each branch of government has its own powers and responsibilities. The executive branch is headed by the President of the United States who carries out federal laws and recommends new ones, directs national defense and foreign policy and performs ceremonial duties. The legislative branch is headed by Congress which consists of the House of Representatives and the Senate. Their main task is to make the laws. The judicial branch is headed by the Supreme Court. This branch interprets the laws, decides cases in which federal laws apply and settles troubles between states.

To prevent one branch from becoming too powerful, the Constitution built in a "check and balance" system. Each branch is controlled by the other two in several ways. The President may veto a law passed by Congress, but Congress may override his/her veto with a two-thirds vote. The Senate must approve any treaty the President makes and approve many of the appointments he/she makes. Any money the President needs for national defense must come from the Congress. The Supreme Court may check the Congress or the President by declaring a law unconstitutional. The Court is appointed by the President, approved by the Senate and may be impeached by Congress.

Label the branches of the government below and write in their Constitutional duties.

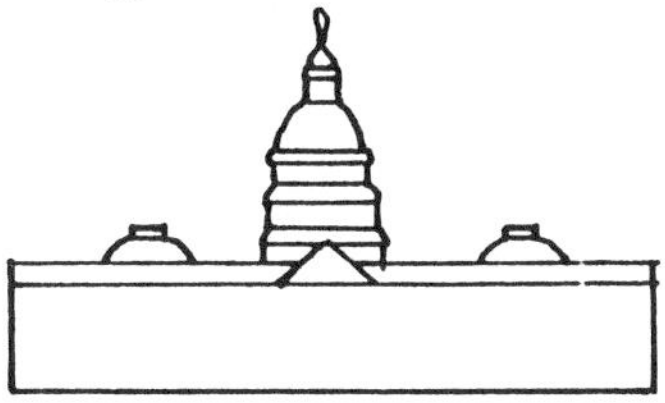

BRANCH

Headed by ____________________

The House of Representatives

One of the two lawmaking bodies established as the result of the Great Compromise was the House of Representatives. It is often just called the House. The Constitution provided for its members, called representatives, to be elected by eligible voters in the states. The House is also called the Lower House because it has always been the lawmaking body elected by the people every two years. This name goes back to when the common man was represented in the Lower House in England. The House is the larger of the two lawmaking bodies of Congress.

The Constitution states the qualifications for a representative. Look in Article I, Section 2, paragraph 2 and tell what the qualifications are. ______________________________

____

____

____

The Constitution says the members of the House of Representatives should select a Speaker from its membership to preside over its meetings. All representatives have legal immunity as members of the House. That means they are protected for anything they write or say while in office. The Constitution also states that all representatives be paid for their service.

The Constitution gave Congress the power to determine the size of the House and to divide the representation in each state according to population. It also provided at least one representative per state. Originally, there was a representative for every 30,000 people in a state. Now, there is one for about every 572,000 citizens. Seven states—Alaska, Delaware, Montana, North Dakota, South Dakota, Vermont and Wyoming—have one representative today. California has the most representatives with 52. A census is required every ten years by the Constitution so the number of representatives may be reapportioned if necessary. When the first House met, 59 members were present. At the end of the first session, there were 65. In 1929, legislation was passed limiting the number of representatives allowed.

Write the names of your states' U.S. representatives below.

____

How much money do these representatives make? __________

____

Do you think this is a lot for the work they do? __________

____

____

____

____

Find a newspaper or magazine article about a U.S. representative. Fill in the information asked for below.

State he/she represents ______________________

Political party he/she belongs to ______________________

Issues he/she is concerned with ______________________

Topic of the article ______________________

Number of years as a representative ______________________

Your opinion of the representative based on the article ______

____

____

Knowing About the House

Use pages 26-27 to help you fill in the blanks below. The boxed letters in the answers will spell out the number of representatives the House is limited to now.

1. _ _ _ _ □ _ _ _ _ _ has more representatives than any other state.

2. The number of representatives allotted per state is based on its _ □ _ _ _ _ _ _ _ _ .

3. A _ _ _ _ □ _ is taken every ten years to keep the population count up-to-date.

4. There is a representative for every _ _ _ _ _ _ _ _ □ _ _ _ _ _ _ _ _ _ -_ _ _ _ □ _ □ _ _ □ _ citizens in a state.

5. One state allotted only one representative is □ _ _ _ _ _ □ _ .

6. The Lower House was so named in □ _ _ _ _ _ □ first because it represented the common man.

7. There were _ _ _ □ _ -_ _ _ _ representatives when the first House met.

8. Originally, there was a representative for every _ _ _ _ _ _ _ □ _ _ _ _ _ _ people in a state.

9. _ _ _ _ □ _ _ is allotted only one _ _ _ □ _ _ _ _ □ _ _ _ _ _ .

10. When the first House session ended, there were _ _ _ _ □-□ _ _ _ representatives present.

11. All representatives have legal ☐ _ _ _ _ _ _ _ .

12. A representative must be _ _ _ _ _ _ - _ _ ☐☐ years old to be elected to the House of Representatives.

The number of members in the House is limited to

_ _ _ _

_ _ _ _ _ _ _

_ _ _ _ _ _ -

_ _ _ _

About half of all nations have a two-house legislature. Write the names of three countries' lower houses that are similar in function to the U.S. House of Representatives. ______________

____

____

List the responsibilities of the Speaker of the House. ______________

____

____

____

____

Explain the role of committees concerning the House. ______________

____

____

____

____

____

The Senate

The other lawmaking body of Congress is the Senate. It is also called the Upper House. Both bodies have about the same amount of power. The Senate can introduce all types of legislation except spending bills. Only the Senate can approve or reject treaties and certain Presidential nominations for government offices.

Originally, the Constitution gave each state legislature the power to select the senators from its state, but that was changed in 1913 by the 17th Amendment. Now, voters in each state elect them. There is equal representation from each state in the Senate. Two senators are elected regardless of the states' population. A senator is elected every six years. No two senators are elected for six-year terms at the same time from one state.

The Constitution states the qualifications for being a senator. Look at Article I, Section 3, paragraph 3 and tell what the qualifications are. ____

____

____

____

The only duty given the Vice President of the United States in the Constitution is that of president of the Senate. The Vice President presides over the sessions of the Senate but may only vote in case of a tie. The Senators choose a president pro tem from their membership to preside over the sessions when the Vice President cannot be there. The Constitution provides all senators legal immunity and compensation for their services.

Only twice have the qualifications of a senator been questioned. Follow the correct path to the Capitol on page 31 to find out what they were.

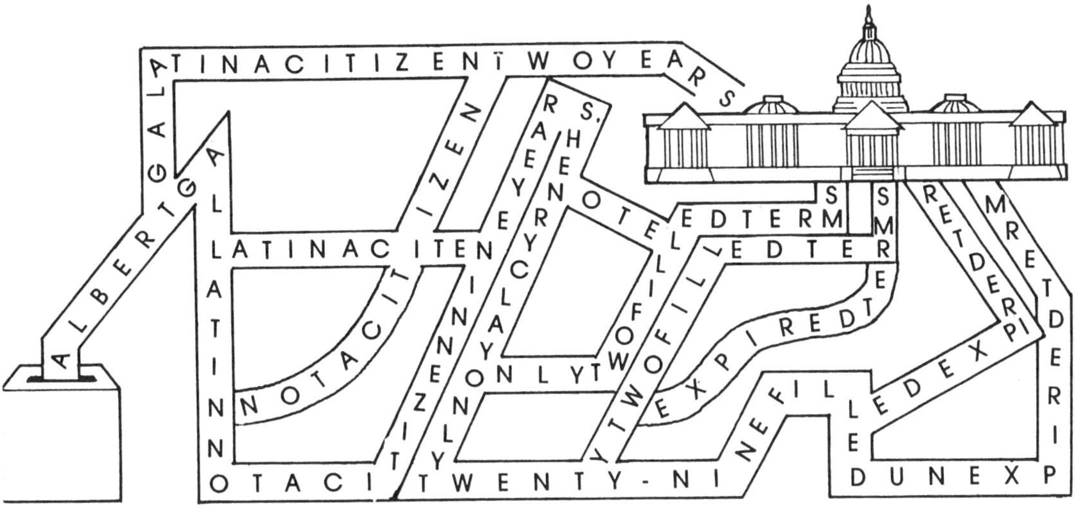

Answer the questions below to learn more about the Senate and its members.

1. How much money do U.S. senators make each year?

   ____

2. What other benefits do senators receive? ______________________________

   ____

   ____

3. Name three other countries that have senates. ______________________________

   ____

4. What role do Senate committees play in the Senate?

   ____

   ____

   ____

5. List the responsibilities of the Senate. ______________________________

   ____

   ____

   ____

   ____

Making Laws

Both senators and representatives may introduce bills, but only members of the House may introduce bills that deal with taxes or spending. Both houses of Congress must pass identical versions of a bill before it can become law. Once a bill is introduced in either house, it goes through almost the same process.

When a bill is (1) introduced, it is (2) assigned to a committee for consideration. If the committee does not think a bill is worthy of further consideration, it (3A) tables it. If the committee thinks otherwise, it (3B) releases it for the entire house's consideration. Most bills are often passed in the House without any opposition. But if there is some disagreement, there can be a debate. The amount of time allowed for the debate is set by its rules committee. In the Senate, the rules allow a senator to speak as long as he/she wants. A senator may choose to filibuster to block a vote, but the Senate may vote cloture to end a filibuster or limit a debate.

When a bill is (4) passed by either house, it is (5) sent to the other for its input and consideration. Any differences between either lawmaking body is worked out by a (6) joint committee. When (7) both houses agree, the bill is signed by the Speaker and the Vice President. Then, it is (8) sent to the President for his/her signature.

Answer the following questions.

Who may introduce a bill? ______________________________

What is the only exception? ______________________________

____

What is the first thing that happens to a bill when it is introduced?

____

If the committee does not like the bill, it may table it. What does that mean? ______________________________

____

Define filibuster. ______________________________

Define cloture. ______________________________

Who sets the time limits for debates? ______________________________

____

Complete the crossword puzzle below.

1. After a bill is signed by the Speaker and the Vice President, it is sent to the _____ .
2. Both houses of Congress must pass _____ versions of a bill before it can become law.
3. When both houses agree on a bill, it is signed by the _____ and the Vice President.
4. A senator may choose to _____ to block a vote.

If you could introduce one bill into Congress, what would it be and why? ______________________________

____

____

____

____

____

____

____

On to the President

Once a bill has been approved by the House and the Senate, it is sent to the President. The President may do one of several things. (9A) He/she may sign it and send it back to the house in which it originated, and it will become law. (9B) If he/she does not approve the bill, he/she may veto it. That is, he/she may return it unsigned to the house in which it originated before listing his/her reasons for disapproval. (B1) Congress may make changes that will meet with the President's wishes, but if it wants the bill to remain as it is, (B2) two-thirds of both houses must vote for it as is, and the bill will become law in spite of the President's veto. (9C) If the President does not act on a bill sent to him/her by Congress within ten working days after receiving it, the bill automatically becomes law. (9D) If Congress adjourns before the ten-day period is up and the President has not signed the bill, the bill does not become law.

Use the information on this page and on page 32 to complete the chart on page 35. Show the different steps and routes a bill may take. Clue: The numbers in the paragraphs will help you.

How a Bill Becomes a Law

1. ______________________________

2. ______________________________

3A. ______________ 3B. ______________

4. ______________________________

5. ______________________________

6. ______________________________

7. ______________________________

8. ______________________________

9A. __________ 9B. __________ 9C. __________ 9D. __________

B1. __________ B2. __________

Impeachment

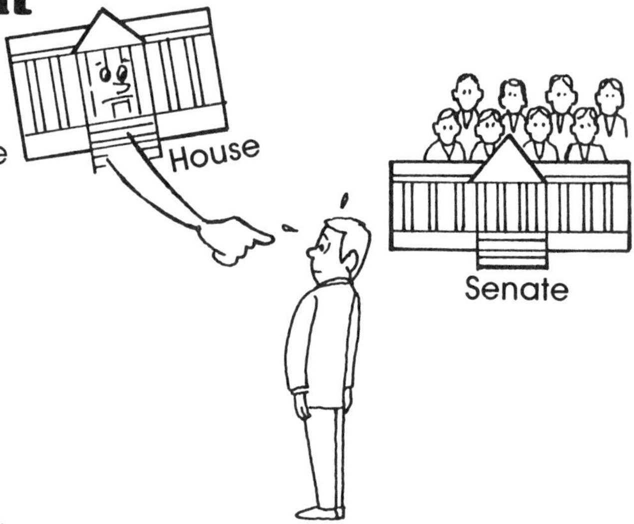

According to Article I, the House of Representatives has the power to impeach all government officials except members of Congress. Each house may punish its own members for disorderly behavior.

Impeachment is the charge of misconduct against an official in office. A majority vote is needed by House members to bring impeachment charges against a government official in office. The Senate has the power to try impeachment cases. A two-thirds vote is necessary for conviction. If an official is found guilty, the official may be removed from office and never allowed to hold a United States government position again. But a conviction does not mean prison or a fine. The case would have to be tried in a regular court for that type of punishment to occur. When a President is tried, the Chief Justice of the Supreme Court presides. The Vice President presides over all other impeachment cases. Why do you think the Vice President does not preside over an impeachment trial if the President is being tried? ______________________________

____

____

The House of Representatives has brought impeachment charges only twelve times. There have been only four convictions. All four were judges. One President, the 17th, was impeached for several crimes while in office. He was found guilty by 35 senators and acquitted by 19—one short of the two-thirds necessary to convict an official. Another President, Richard M. Nixon, resigned before the House voted on articles of impeachment suggested against him by the House Judiciary Committee.

Fill in the answers below and the impeached President's name will appear in the boxes reading down.

1. Who presides over the trial of the President? _ _ _ _ _ ☐ _ _ _ _ _ _
2. Who brings impeachment charges? _ ☐ _ _ _
3. What vote is needed for conviction? _ _ _-_ ☐ _ _ _ _
4. The House may not charge members of _____ . _ _ ☐ _ _ _ _ _
5. Who tries impeachment cases? ☐ _ _ _ _ _
6. How many convictions have there been? _ ☐ _ _
7. How many Presidents have been tried? _ ☐ _

Define impeach. ____

____

____

An impeachment is an accusation. This means that the impeached person may continue to perform the responsibilities of his/her office until he/she has been tried and found guilty of the charges. Write about a situation in which someone is charged with impeachment. It can be real or pretend.

____

____

____

____

____

____

Comparing the Houses

Label the two houses under their pictures. Then, fill in the rest of the blanks correctly.

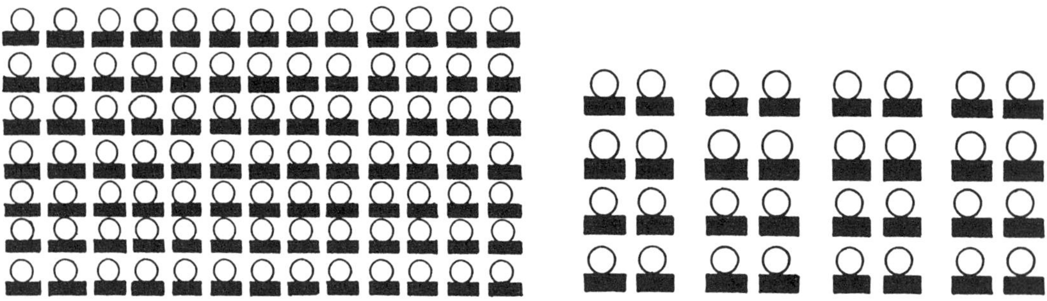

How can you tell which house is which? ______________________________

____

____

____

What is another name for each house? ______________________________

____

These two houses together are called ______________________________.

The members of each house are called ______________________________

______________________________.

Members of both houses receive a ______________________________ for their services.

Members of both houses have ______________________________ for anything they say or write while in office.

Write the qualifications for each house member below.

House of Representatives

Age: ______________________________

Citizenship: ______________________________

Residence: ______________________________

Length of Term: ______________________________

Presiding Officer: ______________________________

Impeachment Power: ______________________________

What vote is necessary to get impeachment? ______________

Introduction of Bills: ______________________________

____

Passage of Bills: ______________________________

____

____

Senate

Age: ______________________________

Citizenship: ______________________________

Residence: ______________________________

Length of Term: ______________________________

Presiding Officer: ______________________________

Impeachment Power: ______________________________

What vote is necessary to get impeachment? ______________

Introduction of Bills: ______________________________

____

Passage of Bills: ______________________________

____

The Executive Branch

The President heads the executive branch of the United States. In what article of the Constitution are the duties of the President stated?

____

In Section 1, paragraph 1 of that article, it states the length of the President's term. What is it? ______________________

Who else is elected at the same time period?

____

What are the three qualifications a person must have to be President? ______________________________

____

____

Write the oath the President of the United States must take before entering the Presidency—Article II, Section 1, paragraph 8. ______________________________

____

____

____

____

Pretend you have to make up a new oath for the President of the United States. Write it below. ______________________________

____

____

____

____

____

____

Section 2 of the article states what the President's duties are. Place a check mark next to the duties given to the President by the Constitution.

___ 1. The President may make treaties by himself/herself.

___ 2. The President appoints the Vice President.

___ 3. The President is commander in chief of the armed forces.

___ 4. The President makes appointments of ambassadors with the approval of the Senate.

___ 5. The President sees that federal laws are carried out as they are designed.

___ 6. The President does not tell Congress what he/she wants.

___ 7. The President has Congress greet visiting ambassadors.

___ 8. The President may make treaties with the approval of the Senate.

___ 9. The President commissions officers in the armed forces.

Answer the questions below to learn more about the executive branch.

1. How much money does the President of the United States make? ______________________________

2. Why do you think this information is made available to the public? ______________________________

   ____

   ____

3. Why might some people not want others to know how much money they make? ______________________________

   ____

   ____

   ____

   ____

The Electoral College

The Electoral College was created by the Constitution because the Founding Fathers did not want the President elected by Congress. This was because he/she would then be under the control of the legislature. They also did not want the President elected by the people. Therefore, the Electoral College, a group of delegates chosen by the voters to elect the President and Vice President, was formed.

On Election Day, the first Tuesday after the first Monday in November, voters mark a ballot for President and candidates, but they select electors, or delegates, to represent their state in the Electoral College. Each state has as many votes in the Electoral College as it has senators and representatives. The electors meet in December on a date set by law to cast their votes. The results are sent to the president of the Senate who opens them. A candidate must receive a majority of the electoral votes to win. If no candidate has a majority, the state delegations in the House of Representatives choose the President from the three candidates which have the highest number of electoral votes. Each state gets only one vote in this type of election.

After two representatives from each body of Congress have counted the electoral votes, the results are officially announced in January. The public knows the results right after the November election because the news media figures them out. They are not official, however, until the electoral votes have been counted by Congress. A candidate may win the popular vote but lose the election.

Answer **True** or **False** to the following statements. The underlined letters in the **True** answers spell out the names of two Presidents who lost the popular vote but won the election.

__________ The electoral results are announced in January.

__________ The Founding Fathers did not want Congress to select the President.

__________ Each state has the same number of electors.

__________ The first Tuesday after the first Monday in November is Election Day.

__________ December is when the Electoral College meets.

__________ Voters actually vote for the delegates.

__________ Congress and the public nominate the candidates.

__________ A ballot is marked by voters in November.

__________ A majority of electoral votes makes a winner.

__________ Each state has as many electors as it has senators and representatives.

__________ The election for President is the first Tuesday in December.

__________ The electors cast their votes in their respective state capitols.

The two Presidents are ______________________________ and ______________________________ .

Presidential Facts:

Name of current President ______________________________

Place of birth ______________________________

Schools attended ______________________________

Political party ______________________________

The Judicial Branch

The Supreme Court heads the judicial branch of the United States government. It is the highest court in the nation and is the only court established by the Constitution. What article in the Constitution states the limits of the Supreme Court? __________

The Supreme Court usually makes decisions of national importance. The Court acts within the laws stated in the Constitution. Because the wording of the Constitution is sometimes hard to understand, it can be difficult to interpret the law. That is one of the duties of the Supreme Court. When the court does make a decision, all other courts in the country must follow that decision to guarantee equal legal justice to all Americans. The Constitution also gives the Supreme Court the power to judge whether federal, state and local governments are acting within the law and also to decide if an action of the President is constitutional.

The Supreme Court has nine members. They consist of a chief justice and eight associate justices. The exact number is set by Congress and, over the years, has changed. Before 1869, the court had only six members. Since that time, it has had nine.

Justices have no set qualifications. However, the Constitution does state that they shall be appointed by the President with the advice and consent of the Senate. Most of them have also been prominent judges, lawyers, law teachers or government officials.

Answer the questions below. The boxed letters in the answers, when unscrambled, will spell out what a judge in the Supreme Court is called.

The Supreme Court usually only hears what kind of cases?

_ _ ☐ _ _ _ _ _

The laws from what document guide the Supreme Court's decisions?

_ _ _   _ _ _ ☐ _ _ _ ☐ _ _ _ _ ' _

The Supreme Court heads what branch of the government?

☐ _ _ _ _ ☐ _ _

What other courts are there in the United States that must follow the decisions of the Supreme Court?

_ _ _ ☐ _ _ _ , _ _ _ _ _ , _ _ ☐ _ _

What is a judge who sits on the Supreme Court bench called?

_ _ _ _ _ _ _

Write the names of the Supreme Court judges in office today.

1. ____________________
2. ____________________
3. ____________________
4. ____________________
5. ____________________
6. ____________________
7. ____________________
8. ____________________
9. ____________________

Answer the questions below to find out more about the judicial branch of government.

Where does the court meet regularly? ____________________

How much money does the chief justice and the associate justices make each year? ____________________

Why do you think the chief justice makes more than the associate justices? ____________________

____

The Relationship Between States and the Federal Government

The Constitution states in Article IV that the rights a citizen has in one state are the same in all the others. All states must honor the laws made in every other state, but Congress may make laws to say how those laws are carried out. This is how the Constitution attempted to keep the United States as one nation instead of separate states. This article also states how new states may be made, and it guarantees every state its own form of government with federal help if called by the governor.

Do you think the rights should be the same in all states? Why or why not? ______________________________

____

____

____

Why should states honor each other's laws? ______________

____

____

____

Think of a law that would pertain only to your state. Write it below giving reasons for its exclusive application. ____________

____

____

____

____

Congress may bring a new state into the nation, but a new state may not be made by dividing an already existing one into two states, nor can two states combine to make one state. Do you think this is a good idea? ________________ Explain your answer.

____

____

____

Why would a state want to have its own government? ________

____

____

____

____

____

____

When might a governor call on the federal government for help?

____

____

____

____

____

____

____

Write about a time your state or another state needed federal assistance. ______________________________________

____

____

____

____

On another sheet of paper, write about a new state you would create if you could. Include name, population, economy, geography, state bird, state tree, state flower, etc.

Amending the Constitution

The Constitution has been a lasting document. Written over 200 years ago, its authors realized times would change so they provided a way the Constitution could be changed when necessary. Which article provides for such changes?

____

What is a change in the Constitution called?

____

A change to the Constitution may be proposed when either two-thirds of Congress or two-thirds of the states request it. To be accepted as part of the Constitution, the proposed amendment must be ratified by three-fourths of the states. It is not easy to make Constitutional changes. Over 9,000 amendments have been proposed over the years, but only 26 have been ratified by three-fourths of the states. Even with popular support from the population at large, such as the Equal Rights Amendment had, ratification is not always assured.

List some suggestions that are currently being considered for Constitutional reform. ____________________

____

____

____

____

____

Which one do you most agree with? Why? ____________________

____

____

Citizens-at-large can have input into making changes by writing to their representatives in Congress. Choose an existing suggestion being considered for Constitutional reform. Write a letter to a representative in Congress telling him/her how you feel about the suggestion and how you would like him/her to vote. Back up your feelings with good reasons.

Suggestion about which you are writing the letter: ______________

____

____

Dear ______________ ,

____

____

____

____

____

____

____

____

____

____

____

____

____

____

____

____

Sincerely,

A Federal Government

The Constitution designed a federal government—a system which divided the powers of the national government and those of the states. The national government only is allowed to make treaties, coin money, maintain an army or navy, engage in war, and place taxes on imports or exports. State governments, on the other hand, provide many services and regulate many activities for the people of a state. These services include maintaining law and order and enforcing criminal law.

Write **N** in front of the powers below and on page 51 if they are controlled by the national government. Write **S** if they are controlled by the states.

____ A tax is placed on products coming to America from foreign countries.

____ Money is given to repair state roads.

____ An agreement is reached between a European country and America.

____ A new stamp is designed to commemorate the signing of the Constitution.

____ Speed limits on rural highways are established.

____ A child must be of a certain age before entering public school.

____ The borders between two states will be a river.

____ The election day date for state officials was set.

____ Money was allotted to build five new aircraft carriers for the United States Navy.

____ Children must attend school until they reach a certain age.

____ The value of money is established.

____ Anyone making fake money or stamps will be punished.

____ A limit was passed on the number of new immigrants allowed to come to America each year.

____ Every driver of an automobile must have a license.

____ There are rules for becoming an American citizen.

____ A citizen will be fined for going through a stop sign.

____ A government official will be tried for selling information to a foreign government.

The federal government has certain constitutional responsibilities toward the states. For example, it cannot divide or break up a state without the state's consent. It is also responsible for protecting the states against invasion and violence, and each state is also guaranteed a republican form of government.

Write about an instance in which the federal government has fulfilled one of the responsibilities listed above.

____

____

____

____

____

If you could make up one law for our country, what would it be and why? ____

____

____

____

____

____

If you could delete one law in our country, what would it be and why? ____

____

____

____

____

Ratifying the Constitution

Because the writers of the Constitution had not been given the authority by their states to write the document, it had to be sent to each state for approval. Before it could be sent to the states, it had to be written in good, easy-to-read form. The delegates asked Gouverneur Morris, an excellent writer, to write it. He did, and in two days, he completed its 4,300 words. On September 17, 1787, thirty-nine of the fifty-five delegates signed it and sent it to the states for ratification. Which article in the Constitution provided the rules for its ratification? ___________

People against the Constitution, called Anti-Federalists, were afraid individual rights were not guaranteed. George Mason, Elbridge Gerry and Patrick Henry spoke out against supporting its passage. They, and many other critics of the document, objected that a bill of rights had not been included, that the President had too much independence, and that the Senate was too aristocratic. Many also believed that Congress had too many powers and the national government had too much authority.

Both the Federalists (supporters of the Constitution) and the Anti-Federalists promoted their causes in newspapers, in pamphlets and in debates in the ratifying conventions. The first American political parties developed from these two groups of people.

Special conventions were held in each state to approve the Constitution. How many states needed to approve the Constitution before it could become law? ___________

Take one side or the other and fight for or against ratification of the Constitution.

List the states in the order in which they ratified the Constitution. Include the dates of ratification.

1. ______ 8. ______
2. ______ 9. ______
3. ______ 10. ______
4. ______ 11. ______
5. ______ 12. ______
6. ______ 13. ______
7. ______

The Bill of Rights

The delegates of the Constitutional Convention had seen no need for a bill of rights because of the limitations set on the federal government by the Constitution and because most of the states had their own bill of rights. But the citizenry felt differently. When the Constitution was sent to the states for ratification, some people would not approve it until there was a bill of rights specifically listing the individual rights of every citizen. Others ratified it with the promise that there would be a bill of rights.

When the first Congress met in 1789, it immediately took several amendments that had been suggested during the ratification process under consideration. James Madison wrote twelve of them which were again presented to the states for ratification. Ten were approved. They make up the Bill of Rights, or the First Ten Amendments to the Constitution. Because they were written and adopted so soon after the original Constitution, they are considered a part of it.

Use a copy of the Bill of Rights to complete the activities below and on page 55.

The 1st Amendment lists five freedoms. Circle them in the box below.

RELIGION	**BREAKING THE LAW**	**SPEECH**
PETITION	**GATHERING PEACEFULLY IN A GROUP**	
TREASON	**HURTING OTHERS**	**PRESS**

Answer the questions below concerning the Bill of Rights.

Which amendment states that a search warrant is necessary before any citizen's possessions may be looked at and seized?

____

Amendments 5 and 6 deal with citizens' rights in courts. Check the ones below that are true.

___ 1. No person may be tried twice for the same crime if found not guilty.

___ 2. Anyone can be tried for a serious crime without a grand jury's indictment (a formal charge).

___ 3. No person must speak against himself/herself.

___ 4. A person may be tried without a jury.

___ 5. Witnesses can be a part of a person's trial.

___ 6. Life, liberty and property cannot be taken without due process of law.

There is much controversy over the 2nd Amendment. Give your opinion of this amendment. ______________________________

More About the Bill of Rights

Answer the questions below to learn more about the Bill of Rights.

Which other amendment provides an army for defense and the right for people to own guns for their protection?

____

Which amendment deals with jury trials?

____

Soldiers can live in private homes in time of war if prescribed by Congress. Otherwise, they may only do so with permission of the owner. Which amendment states this? ____________________

Read the 8th Amendment. What is bail? ____________________

____

____

What three things does the 8th Amendment forbid? __________

____

____

____

The authors of the Bill of Rights could not list every individual right, so they put in the 9th and 10th Amendments to cover those not listed. What might some rights be that were not specifically listed? ____________________

____

____

____

____

____

____

Look at the pictures. Tell which amendment protects the pictured right. If it is not a specific right listed, write **9 and 10**.

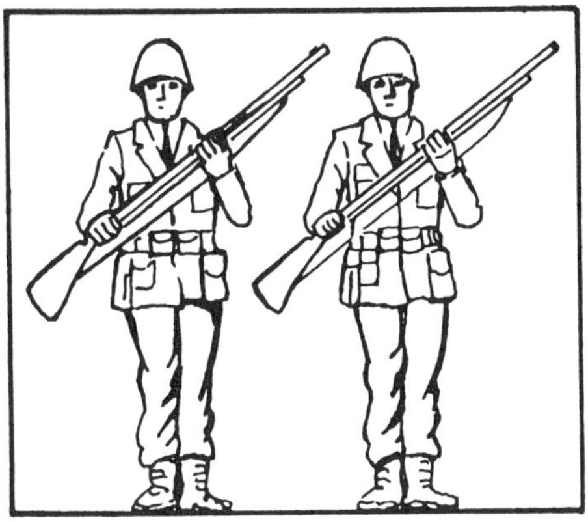

Amendments

Nothing is perfect—and certainly not forever. The Founding Fathers realized this when they provided for changes in the Constitution. Amendments to the Constitution have either been additions or changes to the original document. Since the Bill of Rights was added to the Constitution in 1791, only sixteen amendments have been ratified.

Listed below on the left are the amendments and the dates they became part of the Constitution. On the right and continued on page 59 are descriptions of the amendments in a scrambled order. Read a copy of the amendment section of the Constitution and write the amendment's number on the line in front of its definition.

Amendment 11 (1795)
Amendment 12 (1804)
Amendment 13 (1865)
Amendment 14 (1868)
Amendment 15 (1870)
Amendment 16 (1913)
Amendment 17 (1913)
Amendment 18 (1919)
Amendment 19 (1920)
Amendment 20 (1933)
Amendment 21 (1933)
Amendment 22 (1951)
Amendment 23 (1961)
Amendment 24 (1964)
Amendment 25 (1967)
Amendment 26 (1971)

____ Repealed 18th Amendment but allowed states that wanted to to keep it

____ Described rights of citizens, representation and voting, and defined the obligation of oath takers and Civil War debts

____ Says no one may be kept from voting because of nonpayment of a tax

____ Did away with slavery

____ Gave vote to citizens eighteen and older

____ Gave women the right to vote

____ Limited length of presidential term

____ Changes who elects senators

____ Provided for succession to the presidency and presidential disability

____ Gave everyone the right to vote

____ Changed the dates of the President and Vice President's term in office

____ Would not allow liquor to be made or sold

____ Gave people who live in Washington, D.C. the right to vote in presidential elections

____ Explained what kind of cases federal courts could try

____ Established the income tax

____ Changed how the Electoral College voted

Which amendment do you feel is the most important? Write it below and tell why. ______________________________________

____

____

____

____

____

____

Is there an amendment you think the U.S. doesn't really need anymore? Write about it below. ____________________________

____

____

____

____

____

____

Amendment 11

Amendment 11 changes one part of Article III, Section 2 in the Constitution. Article III, Section 2 tells what cases will be heard by a federal court. Amendment 11 says that it is impossible for a citizen of one state to sue another state in federal court.

Answer the following questions.

Why do you think this amendment was written? ________________

____

____

Write about a circumstance in which a state did something to a citizen of another state that he/she thought was unfair.

____

____

____

____

____

____

If a citizen of one state thought he/she had something against another state, at what level could he/she be heard? __________

____

Suppose you and your family were touring in a neighboring state's capitol building and the ceiling caved in on all of you. Your father received a concussion from the falling cement and the rest of you received bad cuts and bruises. What would you do? ______________________________

____

Explain how you would go about it and what you would claim.

____

____

____

____

____

Amendment 11 resulted from the 1793 case of *Chisholm v. Georgia*. In this case, a man from South Carolina sued the state of Georgia over an inheritance. Georgia argued that it could not be sued in federal court. However, the Supreme Court ruled that it could be. Write about another landmark case involving the Supreme Court below.

____

____

____

____

____

Amendment 11 was proposed on March 4, 1794. It was not ratified until February 7, 1795. Why do you think it took so long to be ratified? Would you have ratified it? Write your answers below.

____

____

____

____

____

____

Amendments 12, 20, 22 and 25

Amendments 12, 20, 22 and 25 deal with the executive branch of the U.S. government.

Amendment 12 changed Article II, Section 1, paragraph 3. It was written after the election of 1800 in which Thomas Jefferson and Aaron Burr tied for President. Because of the tie, the House of Representatives had to select a President. The House took so long to decide before its members finally chose Jefferson, that the people were afraid that Inauguration Day would arrive before a decision was reached. It was at this time that the people decided something had to change.

Before Amendment 12, which was ratified on July 27, 1804, each elector (member of the Electoral College) voted for two men, not saying which he wanted for President. Then, the person with the most votes became President, and the runner-up became Vice President. Amendment 12 changed this.

According to Amendment 12, the electors had to vote for one man for President and another for Vice President. Another change Amendment 12 made was that the House of Representatives would have to elect a President from the three people on the list who received the most votes if no person received a majority. The Constitution had previously said that it should choose from the five highest on the list.

The House of Representatives had to elect a President one other time, in 1825. To find out who it was, answer **True** or **False** to the following statements. The underlined letters in the True answers will spell out that President's name.

________ Amendment 12 changed the date the electors met to elect a President.

________ The Jefferson-Burr tie caused Amendment 12 to be written.

________ The House of Representatives is required to elect a President when there is not a majority for one candidate.

________ Aaron Burr and Thomas Jefferson shared the presidency.

________ Article I deals with the executive branch.

________ Originally, electors did not have to specify which candidate they wanted as President.

________ Amendments 12, 20, 22 and 25 deal with the presidency.

The name of the other President elected by the House of Representatives was ______________________________ .

Rewrite the false statements to make them true.

____

____

____

____

____

____

____

____

____

____

____

Continuation of Amendments 12, 20, 22 and 25

Amendment 12 referred to the fourth day of March as being the starting date of the presidential term. Amendment 20, adopted almost 130 years later, changed the term of the President and Vice President to begin at noon on the twentieth day of January. This was done so there would not be such a long period of time between the election in November and when the President took office. Originally, that much time was needed because transportation was not good, and it took time for a newly-elected President to reach the capital. Now, a President can be in Washington in a matter of hours.

An outgoing President is considered a "lame duck." Before the date change, the country was at a standstill the four months between the election and inauguration dates. This section is the most important part of Amendment 20.

Amendment 20 makes other changes and additions to the Constitution. The second part changes Article 1, Section 4, paragraph 2 of the Constitution. It says that Congress must meet at least once a year and that that meeting is to begin at noon on January third. In the beginning, Congress only met once a year or when the President called a special session. Now they meet almost year-round. The rest of Amendment 20 explains how a President will be chosen if a President-elect dies before taking office and the procedures necessary to get Amendment 20 approved.

Answer the following questions. Use the information on page 64 to help you.

Amendment 20 is often called the Lame Duck Amendment. Why do you think it is called that? ______________________________

____

What is a "lame duck" President? ________________________

____

Why was there such a long period between election day and Inauguration Day? ______________________________________

____

Why is it different now? _____________________________

____

What other change did Amendment 20 make to the Constitution? ____

____

When did Congress meet before? _______________________

What addition did it make to the Constitution? ___________

____

____

Inauguration festivities for the President today are very interesting and have changed greatly over the years. Pretend you were just elected President. Plan your inauguration festivities. Make your day start at 6:30 a.m. and end at midnight. ____________

____

____

____

____

____

____

____

Continuation of Amendments 12, 20, 22 and 25

Amendment 22 was written after Franklin D. Roosevelt was elected to his fourth term. There had been no limit in the Constitution to the number of terms a President could serve. Until Roosevelt's time, no President had served more than two terms. Many felt that this was too long a time for one person to serve, and so the twenty-second Amendment was written. Amendment 22 states that no person may be elected more than twice, and that anyone who has been President for more than two years of someone else's term cannot be elected more than once.

Amendment 25 changes and expands Article II, Section 1. It came about when Lyndon Johnson moved into the Presidency after John Kennedy was assassinated and there was no Vice President for almost fourteen months. Amendment 25 says that when there is no Vice President, the President should nominate one, and if half of the members of the Senate and House of Representatives vote in favor of the nomination, then that person will be the Vice President. Death had been the only way a President had vacated the office until 1974. In 1974, President Nixon resigned. Gerald Ford, who had been appointed Vice President when the vice presidency was vacated earlier, took over the office of President. With the vice presidency again vacated, President Ford nominated Nelson Rockefeller for Vice President. The amendment worked. Amendment 25 also provides for the Vice President to take over the Presidents's duties when the President is unable to perform his job. This part of the amendment proved it worked, too. When President Reagan had surgery in 1985, Vice President Bush assumed the presidency for eight hours.

Answer the following questions. Use the information on page 66 to help you.

Why was Amendment 22 written? ______________________________

____

____

What is the most number of years a President can serve? _______

If a President can only be elected to two terms, then explain how he/she could serve that number of years? ____________________

____

____

Do you agree that the number of years a President can serve should be limited? Why or why not? ______________________

____

____

How long do you think our current President will be in office?

____

Make a list of reasons for your answer. Include as much factual and background information as you can to back up your stance. Debate this issue with a classmate who has taken a position that differs from yours. ____

____

____

____

____

____

____

____

____

____

____

____

____

Continuation of Amendments 12, 20, 22 and 25

Tell under what circumstances the following were President.

Gerald Ford ____

____

George Bush ____

____

Franklin Roosevelt ____

____

Lyndon Johnson ____

____

Richard Nixon ____

____

Review of the Presidential Amendments

Give a title to each of the Presidential Amendments below.

Amendment 12 ____

Amendment 20 ____

Amendment 22 ____

Amendment 25 ____

Which amendment was written after there was a tie for the presidency?____________________

Which amendment was partially written to see that the government was always run by an able person? ______________

Which amendment was written to limit a President's length of service?____________________

Which one redesigned how the President and Vice President were elected?____________________

Which amendment changed the date the President took office?

____

Answer **True** or **False** to the following statements.

_______ A presidency is only vacated upon the President's request.

_______ Electors must specify one candidate for President and another for Vice President.

_______ A lame duck President is disabled.

_______ No President has ever held the office longer than ten years.

_______ One President acted as President for only eight hours.

_______ If the electors fail to give one candidate a majority of the votes for President, the Senate elects a President from one of the three highest vote getters.

_______ When Lyndon Johnson left the vice presidency to become President, there was no Vice President.

_______ Gerald Ford was not elected President.

_______ A person who has been President for more than two years of someone else's term cannot be elected more than once.

_______ In 1974, the United States had, for the first time, both a President and a Vice President who had not been elected to their office.

_______ Before the 25th Amendment went into effect, vacancies in the vice presidency remained unfilled until the next presidential election.

Rewrite the false statements above to make them true.

____

____

____

____

____

____

____

Amendments 13, 14 and 15

Amendments 13, 14 and 15 are often called the Civil Rights Amendments. The Founding Fathers had so many disagreements about the treatment of slaves, that there almost was no Constitution. Slavery by name was not mentioned in the Constitution. Not until after the Civil War was slavery abolished. Amendment 13, ratified on December 6, 1865, gave black people their freedom.

The first section of Amendment 14, ratified on July 9, 1868, made black people citizens of both the United States and of the state in which they lived. It also entitled them to the same rights guaranteed all citizens by the Constitution. Section 2 of this amendment changed Article 1, Section 2 of the Constitution. Every black is now counted as one person instead of three-fifths of a person. This section also imposes a penalty on any state that refuses to let all male citizens, twenty-one or over, vote. The rest of Amendment 14 controlled the positions federal officers who had joined the Confederacy might have had and payment of the Union's Civil War debt.

Whereas Amendment 14 gives all adult males that are citizens the right to vote, Amendment 15, ratified on February 3, 1870, carries it a step further and states that no citizen may be kept from voting because of race, color or previous condition of servitude (slavery).

These three amendments were just the beginning of many advancements and accomplishments for black Americans. Since this time, many acts, laws and other amendments have been passed to ensure that all people enjoy equal rights.

Solve the crossword puzzle below to learn more about Amendments 13, 14 and 15.

Across

1. What was fought between the states?
3. Word not mentioned in the Constitution
4. What Amendment 13 gave all people
6. The amendment that made black Americans citizens

Down

1. What Amendments 13, 14 and 15 are often called
2. No citizen can be kept from voting because of this.
5. Every black person is now counted as _____ person.

Do you think all people should have equal rights regardless of age, race, sex and religion? Why? ______________________________

Amendment 16

Amendment 16 gives Congress the right to tax individuals according to the amount of money they earn. It was first proposed as an amendment in 1894, but it was ruled unconstitutional by the Supreme Court. It became part of the Constitution in 1913. Until then, the only way Congress got money to pay debts and provide the necessary defense for the country was through taxing imported items, and items made, sold and used in the United States. Such taxes were the same in all the states.

The income tax, created by Amendment 16, raised much more money. When the amendment became law, a part of every person's salary was taken out for taxes. It was thought to be a fair tax because it was a way to give the government money for running the country according to each individual's ability to pay. People probably complain more about Amendment 16 than any other amendment.

Do you think Amendment 16, the Income Tax Amendment, is a fair one? Explain your answer. ______________________________

____

____

Think of some ways the government spends the money received from individuals' incomes. Write them below. ________________

____

____

____

Why do you think people complain about Amendment 16? ____

____

____

____

Take a poll. Ask several working people if they think paying an income tax is necessary, if it is fair, and if they can suggest other ways for the government to raise money.

#	NECESSARY?	FAIR?	OTHER WAYS
1			
2			
3			
4			

Pretend you are an adult and that you have your dream job. Explain what kind of job it is and what you do. ____________________

____

Write how much money you make in a year and each month. (Be realistic. Use the necessary resources to help you with the amount if you don't know how much the type of dream job you have pays.) year __________ month __________

From the amount of money you make, deduct the correct amount of income taxes you would have to pay to the federal government each month and year.
month __________ year __________

Now find out how much your state income tax is. Deduct this amount from the amount of money you make each month and year. month __________ year __________

Is any more money usually taken out of people's paychecks each month? If so, write what it is taken out for and deduct a similar amount from the amount you make. ____________________

____

____

About how much money will you actually "take home" at the end of each month/year? month __________ year __________

Do you think this is fair? Why or why not? ____________________

____

____

Amendments 18 and 21

Amendment 18 is called the Prohibition Amendment. This amendment made it illegal to make, sell or ship any alcoholic beverages anywhere in the United States. It also prohibited exporting them out of or importing them into the country. Businesses that made or sold liquor had one year from the time Amendment 18 was passed to stop making it. A time limit of seven years for ratification was placed on this amendment. Since then, most amendments have given a seven year time limit from the time of proposal to ratification. One year after this amendment was ratified, it became law. Fourteen years later, Amendment 18 was repealed by Amendment 21. Amendment 21 withdrew the federal limitations on the manufacturing and selling of alcoholic beverages, but recognized the rights of individual states to still practice prohibition.

Define prohibition. ______________________________

____

Define repeal. ______________________________

____

Do you think a seven year limit for ratification is a good idea? Why or why not? ______________________________

____

____

____

Why did it take one year after ratification for Amendment 18 to become effective? ______________________________

____

____

Did Amendment 21 make prohibition illegal? __________
Explain your answer. ______________________________

____

____

Do you think laws should be made because they are best for the country or because they involve strong personal beliefs and morals? Explain your answer. ______________________________

____

____

____

____

Pretend prohibition has just been put into effect in your community. Make a list of all the effects this would have. Take into consideration the following: businesses, jobs, entertainment, accidents, etc. Write about the effects below. ______________

____

____

____

____

____

____

____

____

____

____

____

Amendments 17, 19, 23, 24 and 26

Amendments 17,19, 23, 24 and 26 all have something to do with voting. They either changed parts of or expanded the Constitution. Amendment 17 changed Article I, Section 3 of the Constitution. Amendment 17 gives the people the right to elect their senators instead of the state legislators. The 17th Amendment also directed the governor of a state to choose a vacated senatorial seat until the election could be held.

Answer the questions below to learn more about Amendments 17, 19, 23, 24 and 26.

How were senators elected before Amendment 17? ____________

____

because the authors of the Constitution thought the general public would not be well enough informed to make the right choice.

How were vacancies filled before Amendment 17? ____________

____

Amendment 19 gave women the right to vote. This cancelled a part of Amendment 14. Explain how it did that. ________________

____

____

____

____

____

People living in the District of Columbia, the United States Capital, could not vote in Presidential elections because it was not a state. Amendment 23 gave them that right. Now, the capital has one elector who meets with the electors from all the states to elect the President and Vice President.

Why didn't the people in the country's capital vote for President before Amendment 23 became effective? ____________________

____

Do you think this was fair? Why or why not? ____________________

____

____

Amendment 24 ruled that no United States citizens could be kept from voting in a federal election because they had failed to pay taxes. Some states tried to keep some people from voting if they had not paid their taxes. Why do you think some states tried this?

____

Do you think this is fair? Why or why not? ____________________

____

____

What do you think should be done to people who do not pay their taxes? ____________________________________

____

Amendment 26 changed the age at which people could vote to 18. Although the Constitution said the age at which people could vote should be set by the states, one amendment stated the age. Which amendment was it and what age did it state?

____

Do you think 18-year-olds should be allowed to vote? Why or why not? ____________________________________

____

Do you think you should be allowed to vote? Why or why not?

____

____

3. 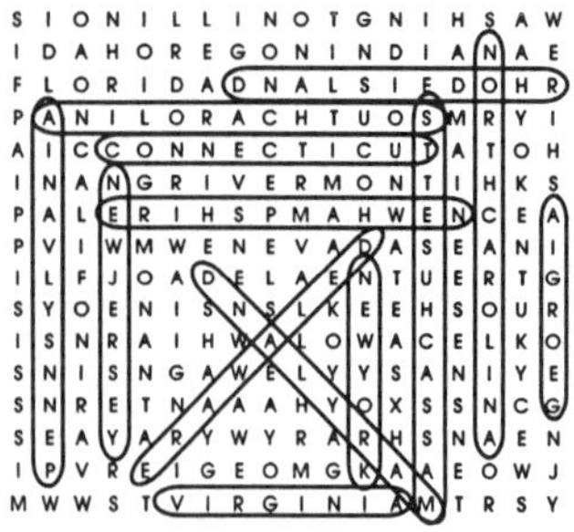

5. 1) George; 2) ten; 3) Boston; 4) imported; 5) England; 6) first; 7) Stamp Act; Georgia

7. 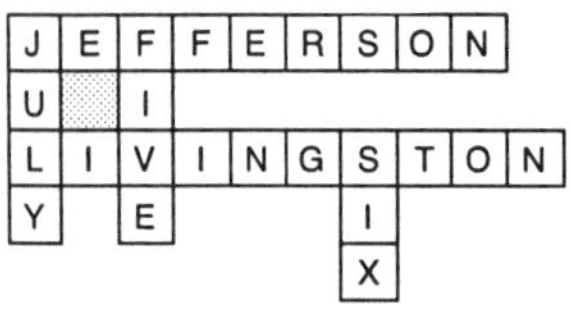

fifty-six; Answers will vary.

9. F; T; F; F; T; F; F; T; F; F; T; F

11. Annapolis; Philadelphia; four months; trade and boundary; Rhode Island

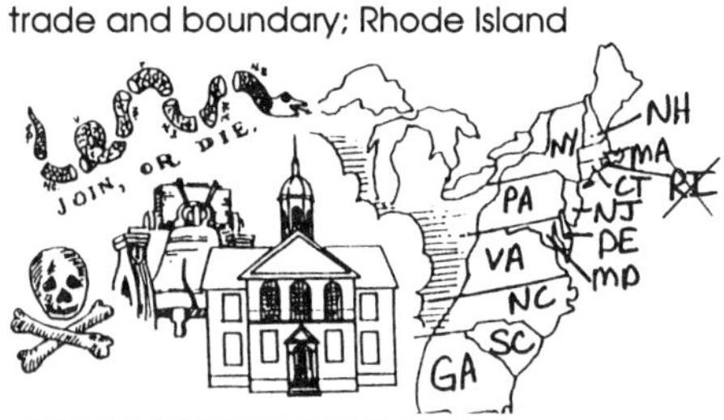

13. 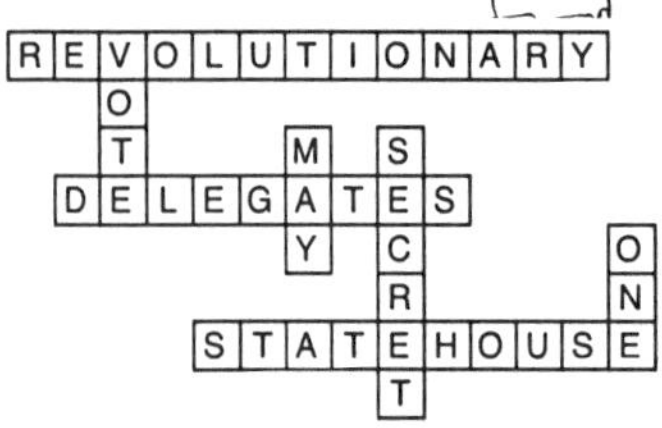

14. disagreement; equal; compromise; several; matter

15. 

1) small; 2) Virginia; 3) Roger; 4) Sherman; 5) New Jersey; 6) Senate; 7) large; Edmund Randolph; William Paterson

16. Although answers will vary, the following suggestions are given. A settlement of differences in which each side makes concessions; The Southern states would get more congressmen if slaves were counted; The Northern states wanted them counted because the states with slaves would have to pay more taxes; Article I, Section 3; The South did not pay as high taxes as the North would have liked. However, it did not get as large a representation as it wanted.

17. Answers will vary.

19. NH—all; NY—Hamilton; MA—Gorham, King; NJ—Brearley, Paterson, Livingston, Dayton; CT—Johnson, Sherman; GA—Few, Baldwin; VA—Washington, Blair, Madison; NC—Spaight, Blount, Williamson; DE—all; SC—all; MD—McHenry, St. Thomas Jenifer, Carroll; PA—all; 39; 40; PA, DE, NH, SC; NY; 7; VA, NY; PA; NH; Both entire delegations signed.

20. T; F; T; F

21. T; T; F; F; F; F; T; F; F; T; All the delegates weren't always present. However, they all contributed their ideas to it; The delegates were educated and experienced; The purpose of the Constitution was to establish a stronger central government; All delegates contributed ideas; Delegates wrote and rewrote the Constitution; All delegates had held at least one public office and all had backgrounds of some financial success; Benjamin Franklin was the oldest delegate; Fifty-five men wrote the Constitution.

22. 3; Preamble, articles, amendments; To introduce and explain the reasons for writing the document; 7; III; VI; I; VII; ratification; II; V; amending; IV; 26; The Bill of Rights and Amendments have kept it up to date.

23. V; Sections of the Constitution defining different powers; Change in the Constitution to reflect changing needs; To pass and accept the laws proposed in the Constitution

25. Legislative: Congress—House of Representatives, Senate, make the laws; Executive: President—carries out federal laws, recommends new laws, directs

national defense and foreign policy, performs ceremonial duties; Judicial: Supreme Court—interprets the laws, decides cases in which federal laws apply, settles troubles between states

26. Must be at least 25 years of age, a U.S. citizen for at least 7 years, must live in state from which chosen

28. 1) California; 2) population; 3) census; 4) five hundred seventy-two thousand; 5) Delaware; 6) England; 7) fifty-nine; 8) thirty thousand; 9) Wyoming, representative; 10) sixty-five

29. 11) immunity; 12) twenty-five; four hundred thirty-five; France, Great Britain, Italy, Germany; The Speaker is the recognized leader of his/her political party in the House. He/she is also the presiding officer and is expected to use the office to promote his/her political party. The Speaker ranks next in line after the Vice President to become President. He/she gives representatives permission to debate; The House has 22 standing (permanent) committees. They do most of the work of preparing legislation. In both houses, nearly every bill goes before a standing committee for action.

30. Must be at least 40 years of age, a U.S. citizen for at least 9 years, must live in state from which chosen

31. Albert Gallatin not a citizen nine years. Henry Clay only twenty-nine, filled unexpired term. 1) Answers will vary. 2) They receive free office space, free health care and extra money to hire committee assistants and pay for such expenses as telephone service and travel. 3) France, Great Britain, Italy 4) Senate committees consist of members of both political parties. They do most of the actual work of the Senate. They deal with bills concerning specific legislative subjects. 5) considers legislation, approves/rejects certain presidential appointments, approves/rejects treaties, elects Vice President if the Electoral College fails to give any candidate a majority, judges impeachment cases brought against U.S. government officials by the House

32. senators and representatives; only members of the House may introduce bills dealing with taxes or spending; It is assigned to a committee for consideration.

33. Remove it from consideration; filibuster—a long talk to obstruct passing a bill; cloture—A vote placing a time limit on a speaker; the rules committee unless a vote is taken in the Senate; 1) President; 2) identical; 3) speaker; 4) filibuster

35. 1) A bill is introduced. 2) It is assigned to a committee for consideration. 3A) Table it. 3B) Release it for House's consideration. 4) A bill is passed by either House. 5) It is sent to the other for consideration and input. 6) Differences are taken to a joint conference committee. 7) Bill is signed by Speaker and Vice President. 8) Bill is sent to the President for signature. 9A) President signs bill and it becomes law. 9B) President vetoes bill. B1) Congress may make changes for President's approval. B2) Two-thirds of both Houses can override the veto. 9C) Bill becomes law after ten working days if President does not act on it. 9D) Bill does not become law if Congress adjourns before ten days are up.

36. He/she might convict the President to become President himself/herself.

37. 1) chief justice; 2) House; 3) two-thirds; 4) Congress; 5) Senate; 6) four; 7) one; impeach—To make an accusation against

38. House of Representatives is on the left. Senate is on the right; The House is a much larger body. The Senate only has two members for each state; Lower House, Upper House; Congress; representatives, senators; salary; legal immunity

39. Representative: 25 years; 7 years; In state that elects him/her; 2 years; speaker; brings charges; majority; can introduce any legislation; majority vote for identical bill as Senate
Senator: 30 years; 9 years; in state that elects him/her; 6 years; Vice President; tries person; two-thirds; can introduce any legislation but money; majority vote for identical bill as House

40. Article II; 4 years; Vice President; be a natural-born citizen, 35 years old, a U.S. resident for 14 years; "I do solemnly swear (or affirm) that I will faithfully execute the Office of the President of the United States, and will to the best of my ability, preserve, protect and defend the Constitution of the United States."

41. Numbers 3, 4, 8 and 9 should have a check mark next to them.

43. T; T; F; T; T; T; F; T; T; T; F; T; Rutherford B. Hayes and Benjamin Harrison

44. Article III

45. national; the Constitution's; judicial; federal, state, local; justice; The Court meets regularly in the Capitol.

48. V; amendment

50. N; S; N; N; S; S; N; S; N; S; N; N; N

51. S; N; S; N

52. VII; 9

53. 1) DE—Dec. 7, 1787; 2) PA—Dec. 12, 1787; 3) NJ—Dec. 18, 1787; 4) GA—Jan. 2, 1788; 5) CT—Jan. 9, 1788; 6) MA—Feb. 6, 1788; 7) MD—April 28, 1788; 8) SC—May 23, 1788; 9) NH—June 21, 1788; 10) VA—June 25, 1788; 11) NY—July 26, 1788; 12) NC—Nov. 21, 1789; 13) RI—May 29, 1790

54. The following should be circled: religion, press, speech, gathering peacefully in a group, petition

55. Amendment 4; Numbers 1, 2, 5 and 6 are true.

56. Amendment 2; Amendment 7; Amendment 3; bail—amount of money court requires accused to put up before being set free; no high bail, no high fines, no cruel or unusual punishments; who goes to school, when cars need inspections, hunting season dates, how many dogs an individual can own, etc.

57. left to right, top to bottom: Amendment 2; Amendment 1; Amendments 9 and 10; Amendment 1; Amendments 9 and 10; Amendment 6

58. 21; 14; 24; 13; 26; 19

59. 22; 17; 25; 15; 20; 18; 23; 11; 16; 12

60. at a state level court

63. F; T; T; F; F; T; F; John Quincy Adams; Amendment 12 changed the procedure of how the President and Vice President are elected; They tied for President; Amendments 12, 20, 22 and 25 deal with the executive branch.

65. Answers will vary; one who continues to serve but has no power because he/she is on the way out; It took longer to complete the count and for Presidents to reach Washington; We have better technology and transportation; The meeting day of Congress; December—the first Monday; How a President will be chosen if a President-elect dies before taking office

67. Lyndon Johnson created a vacancy in the Vice Presidency. People worried about the open slot; 10; If he/she served out two years or less of another President's term and then was elected to two terms

68. Nixon resigned; acting President when Reagan had surgery; elected 4 times; after Kennedy was shot; elected 2 times; 12—electing a President and a Vice President; 20—Lame Duck; 22—length of service; 25—presidential succession and disability; 12; 25; 22; 12; 20

69. F; T; F; F; T; F; T; T; T; T; T; T; Many things can happen to cause a President to vacate from office (i.e. death, resignation); A lame duck President is an outgoing President. He/she continues to serve but has no power; Franklin D. Roosevelt was elected to four terms; The House of Representatives, not the Senate, elects a President from one of the three people who receive the most votes.

71.

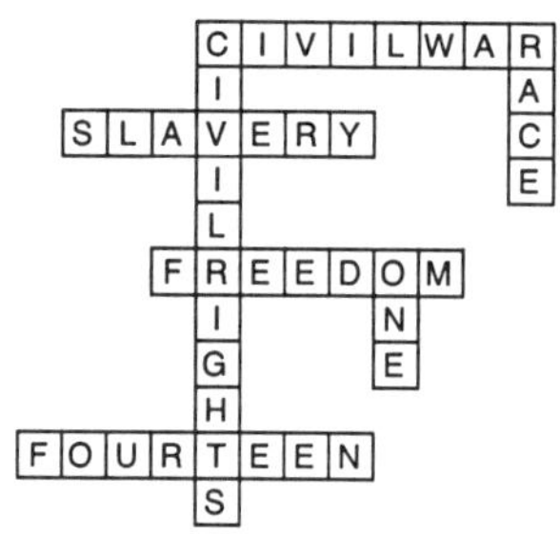

74. To forbid by law the making and selling of liquor; cancel, take back, revoke

75. Businesses had to stop making liquor and get rid of the stock; no, states had the right to continue Prohibition if they wanted

76. by their state legislatures; Governor made temporary appointment until state legislature could; Amendment 14 referred to male voters.

77. It wasn't a state; Answers will vary; to keep poor blacks from voting; Answers will vary; Answers will vary; Amendment 14, age 21; Answers will vary.

